THE MILITARY QUOTATION BOOK

Other quotation books edited by James Charlton

THE WRITER'S QUOTATION BOOK
THE EXECUTIVE'S QUOTATION BOOK
LEGAL BRIEFS

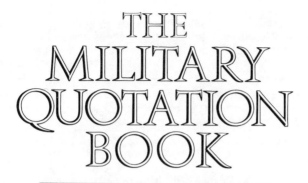

THE MILITARY QUOTATION BOOK

More than 600 of the Best Quotations
about War, Courage, Combat,
Victory, and Defeat

EDITED BY
JAMES CHARLTON

ST. MARTIN'S PRESS
NEW YORK

Design by Mary Kornblum/Hudson Studio

Library of Congress Cataloging-in-Publication Data

The Military quotation book / James Charlton, ed.
p.   cm.
ISBN 0-312-04350-3
1. Military art and science—Quotations, maxims, etc.
I. Charlton, James.
U102.M598   1990   355—dc20   89-77925

10 9 8 7 6

INTRODUCTION

This book is about war. War in all its manifestations and effects: the rattling sabres, the feverish preparations, the battles on land, in the air, and at sea, and the feelings at home. It is about death and impending death, the vanquished and the victorious. And I hope, finally, it is a book about peace.

Some veterans feel that war provided their finest hours and proudest memories; it was the time they felt most alive, when their senses were the keenest. For many, their strongest lifetime friendships were forged with those with whom they served. Nostalgia for one's youth plays a role here, but this is only a partial explanation. To be at war, no matter where one is serving, is to sense palpably the possibility of death; if not to you, then to a relative or friend. If one is a soldier, sailor, or airman, then the likelihood of death is that much more present. British poet Wilfred Owen, a frontline commander in World

War I, complimented one of his men, "Well done. You are doing that very well, my boy." These were his last words, but his poetry lives on.

> What passing-bells for these who die as cattle?
> Only the monstrous anger of the guns.
> Only the stuttering rifles rapid rattle
> Can patter out their hasty orisons.
>
> from *Anthem for Doomed Youth*, by Wilfred Owen

Every war has its pacifists, those who are against the violence no matter what the cause. These are often people who have not served in the armed forces. But there is another rank of pacifists, a more knowing and, to my mind, more persuasive type— those who have seen the face of war up close and are horrified by it. Often these are ex-officers. Dwight Eisenhower, underrated as a commander and as a president, sounded many warnings about the dangers of war and military preparation, and several of his most eloquent criticisms and statements are in this book. His feelings on the subject express well those of many ex-servicemen.

> Every gun that is made, every warship launched, every rocket fired, signifies a theft from those who hunger and are not fed, those who are cold and are not clothed. The world in arms is not spending money alone. It is spending the sweat of its laborers, the genius of its scientists, the hopes of its children. This is not the way of life at all in any

true sense. Under the cloud of war, it is humanity hanging on a cross of iron.

Wars have changed over the centuries, especially in the last two hundred years. Man has gotten very good at being able to kill masses of people, and to do it less personally. "Absolute violence," as Clausewitz called it, has made a war with just a few hundred killed a thing of the past. Now we are much more efficient. Maybe this is a good thing. Maybe we have finally realized how destructive the modern machinery of war is and this *will* act as a deterrent.

But the subject of war will always raise passions and emotions. As a writer, I have collected and edited four books of quotations; when I was an editor-in-chief of a publishing house I published more than a dozen books of quotations, ranging from sports and the Beatles to famous last words. None of those books aroused the level of feelings reflected in this collection.

This book is by no means all-inclusive. Encyclopedic works certainly have their place on the library shelves, but to my way of thinking, a carefully selected anthology is preferable. A shorter book of quotations must take a position; the editor must select and edit, rather than merely gather as many sayings as possible.

I feel a collection of quotations should be a book to read and to browse through, a volume to inspire, amuse, and arouse the reader. A quotation, to be useful, must reflect a point of view, even if it is one that the reader opposes. It must have been said by some authority or someone well known; a wonderfully pithy statement made by your sister is just that. The same line uttered by a prime minister, even if penned by a speech writer, is the one remembered. And quoted. So it is here. I have tried to include only quotations from individuals that any well-read lay person would have heard of. There are some exceptions; either the quote was unique or outstanding or fit the situation perfectly, or the person quoted was a military officer with whom a military reader would be familiar. I hope the reader is pleased with the results.

Many people helped cull through the thousands of available quotations. Captain Gil Diaz, who was with the history department of the United States Naval Academy when this project began, deserves a special note of thanks for his assistance. Cecilia Oh, who was indefatigable in her research and fact checking, was involved in every phase of the project. Barbara Binswanger, Kevin Charlton, and Cindy Behrman helped immeasurably with the editing process. And finally, Barbara Anderson, senior editor at St. Martin's Press, was as professional, supportive, and patient as any author could wish.

Jim Charlton
New York City

War makes rattling good history; but Peace is poor reading.

THOMAS HARDY

Above all, this book is not concerned with Poetry.
The subject of it is War, and the Pity of War.
The Poetry is in the pity.

WILFRED OWEN

Wars make for better reading than peace does.

A.J.P. TAYLOR

Man lives by habits indeed, but what he lives for is thrill and excitements. . . . From time immemorial war has been . . . the supremely thrilling excitement.

WILLIAM JAMES

If peace . . . only had the music and pagaentry of war, there'd be no wars.

SOPHIE KERR

Men grow tired of sleep, love, singing and dancing sooner than war.

HOMER

The Lord is a man of war.

Exodus 15:3

God hates violence. He has ordained that all men fairly possess their property, not seize it.
EURIPIDES

Every man thinks God is on his side.
JEAN ANOUILH

All the gods are dead except the god of war.
ELDRIDGE CLEAVER

Fortune is always on the side of the biggest battalions.
MARQUISE DE SÉVIGNÉ

God is always on the side of the big battalions.
MARSHAL HENRI DE LA TOUR D'AUVERGNE
TURENNE

Oho, the Pope! How many battalions does he have?
JOSEPH STALIN

What mean and cruel things men do for the love of God.
W. SOMERSET MAUGHAM

There are no atheists in fox holes.
WILLIAM THOMAS CUMMINGS

The bombs in Vietnam explode at home; they destroy the hopes and possibilities for a decent America.

MARTIN LUTHER KING, JR.

Vietnam is a military problem. Vietnam is a political problem; and as the war goes on it has become more clearly a moral problem.

EUGENE MCCARTHY

You will kill ten of our men and we will kill one of yours, and in the end it will be you who tire of it.

HO CHI MINH

In the final analysis, it is their war. They are the ones who have to win it or lose it. We can help them, we can give them equipment, we can send our men out there as advisers, but they have to win it, the people of Vietnam.

JOHN F. KENNEDY

What the United States wants for South Vietnam is not the important thing. What North Vietnam wants for South Vietnam is not the important thing. What is important is what the people of South Vietnam want for South Vietnam.

RICHARD M. NIXON

Curtis LeMay wants to bomb Hanoi and Haiphong. You know how he likes to go around bombing.

LYNDON B. JOHNSON

It's silly talking about how many years we will have to spend in the jungles of Vietnam when we could pave the whole country and put parking stripes on it and still be home by Christmas.

RONALD REAGAN

I could have ended the war in a month. I could have made North Vietnam look like a mud puddle.

BARRY GOLDWATER

Within the soul of each Vietnam Veteran there is probably something that says, "bad war, good soldier."

MAX CLELAND

To win in Vietnam, we will have to exterminate a nation.

DR. BENJAMIN SPOCK

The war in Vietnam was bad for America because it was a bad war, as all wars are bad if they consist of rich boys fighting poor boys when the rich boys have an advantage in the weapons.
    NORMAN MAILER

Vietnam was the first war ever fought without any censorship. Without censorship, things can get terribly confused in the public mind.
    GENERAL WILLIAM WESTMORELAND

The publishers have to understand that we're never more than a miscalculation away from war and that there are things we're doing that we just can't talk about.
    JOHN F. KENNEDY

You furnish the pictures and I'll furnish the war.
    WILLIAM RANDOLPH HEARST, *in a telegram to Frederic Remington*

German prisoners, asked to assess their various enemies, have said that the British attacked singing, and the French attacked shouting, but that the Americans attacked in silence. They liked better the men who attacked singing or shouting than the grimly silent men who kept coming on stubbornly without a sound.
    JAMES JONES

Television brought the brutality of war into the comfort of the living room. Vietnam was lost in the living rooms of America—not on the battlefields of Vietnam.

MARSHALL MCLUHAN

American boys should not be seen dying on the nightly news. Wars should be over in three days or less, or before Congress invokes the War Powers Resolution. Victory must be assured in advance. And the American people must be all for it from the outset.

EVAN THOMAS

America is the only nation in history which miraculously has gone directly from barbarism to degeneration without the usual interval of civilization.

GEORGES CLEMENCEAU

Patriotism is easy to understand in America. It means looking out for yourself by looking out for your country.

CALVIN COOLIDGE

We fight not to enslave, but to set a country free, and to make room upon the earth for honest man to live in.

THOMAS PAINE

If the American Revolution had produced nothing but the Declaration of Independence, it would have been worthwhile.

SAMUEL ELIOT MORISON

Let every nation know, whether it wishes us well or ill, that we shall pay any price, bear any burden, meet any hardship, support any friend, oppose any foe, in order to assure the survival and the success of liberty.

JOHN F. KENNEDY

This will remain the land of the free only so long as it is the home of the brave.
ELMER DAVIS

The trouble is not that the Soviets and Americans do not have the same positions; the trouble is that they do not have them at the same time.
LAWRENCE D. WEILER

The Soviet Union can choose either confrontation or cooperation. The United States is adequately prepared to meet either choice. We would prefer cooperation.
JIMMY CARTER

History tells me that when the Russians come to a country, they don't go back.
MOHAMMED DAOUD, *Afghani rebel leader*

I cannot forecast to you the action of Russia. It is a riddle wrapped in a mystery inside of an enigma.
WINSTON CHURCHILL

Somebody said the second most stupid thing in the world a man could say was that he could understand the Russians. I've often wondered what in the hell was the first.
RONALD REAGAN

The enemies of freedom do not argue; they shout and they shoot.
WILLIAM RALPH INGE

An opinion can be argued with; a conviction is best shot.
T.E. LAWRENCE

When you see a rattlesnake poised to strike, you do not wait until he has struck before you crush him.
FRANKLIN D. ROOSEVELT

My men and I have decided that our boss, the president of the United States, is as tough as woodpecker lips.
COLONEL CHARLES BECKWITH

We have met the enemy, and they are ours.
COMMODORE OLIVER PERRY

We have met the enemy, and he is us.
WALT KELLY

We met the enemy and he was us.
GENERAL WILLIAM WESTMORELAND

We used to wonder where war lived, what it was that made it so vile. And now we realize that we know where it lives, that it is inside ourselves.
ALBERT CAMUS

If we see that Germany is winning the war we ought to help Russia, and if Russia is winning we ought to help Germany, and in that way let them kill as many as possible.

HARRY TRUMAN

At Victoria Station the R.T.O. gave me a travel warrant, a white feather and a picture of Hitler marked "This is your enemy." I searched every compartment but he wasn't on the train.

SPIKE MILLIGAN

I do not consider Hitler to be as bad as he is depicted. He is showing an ability that is amazing and he seems to be gaining his victories without much bloodshed.

MAHATMA GANDHI, *May 1940*

Has there ever been danger of war between Germany and ourselves, members of the same Teutonic race? Never has it even been imagined.

ANDREW CARNEGIE

The Nazis are the enemy, Wade into them. Spill *their* blood. Shoot *them* in the belly. When you put your hand into a bunch of goo that a moment before was your best friend's face, you'll know what to do.

GEORGE C. SCOTT, *as General George Patton in* Patton

If men are mad enough they will fight. If not, the ordinary means of diplomacy will do.
WILLIAM GRAHAM SUMNER

Jaw-jaw is better than war-war.
HAROLD MACMILLAN

It is better to have less thunder in the mouth and more lightning in the hand.
GENERAL BEN CHIDLAW

What you do speaks so loud that I cannot hear what you say.
RALPH WALDO EMERSON

The great questions of the day will be decided not by speeches and majority votes . . . but by iron and blood.
OTTO VON BISMARCK

The field of combat was a long, narrow, green baize covered table. The weapons were words.
ADMIRAL C. TURNER JOY

A real diplomat is one who can cut his neighbor's throat without having his neighbor notice it.
TRYGVE LIE

There are a few ironclad rules of diplomacy but to one there is no exception. When an official reports that talks were useful, it can safely be concluded that nothing was accomplished.

JOHN KENNETH GALBRAITH

I'm convinced there's a small room in the attic of the Foreign Office where future diplomats are taught to stammer.

PETER USTINOV

At vast expense the Ambassadors offer up their livers almost every night in the service of their country.

PATRICK O'DONOVAN

The ability to get to the verge without getting into the war is the necessary art. If you try to run away from it, if you are scared to go to the brink, you are lost.

JOHN FOSTER DULLES

The supreme excellence is not to win a hundred victories in a hundred battles. The supreme excellence is to subdue the armies of your enemies without even having to fight them.

SUN TZU

More history's made by secret handshakes than by battles, bills, and proclamations.
  JOHN BARTH

He lied, I knew he lied and he knew I lied. That was diplomacy.
  ADMIRAL WILLIAM KIMBALL

Sincere diplomacy is no more possible than dry water or wooden iron.
  JOSEPH STALIN

It is useless to delude ourselves. All the restrictions, all the international agreements made during peacetime are fated to be swept away like dried leaves on the winds of war.
  GENERAL GIULIO DOUHET

Treaties are like roses and young girls. They last while they last.
  CHARLES DE GAULLE

Negotiation in the classic diplomatic sense assumes parties more anxious to agree than to disagree.
  DEAN ACHESON

The principle of give and take is the principle of diplomacy—give one and get ten.
  MARK TWAIN

No country can act wisely simultaneously in every part of the globe at every moment of time.
HENRY KISSINGER

The time has come to stop beating our heads against stone walls under the illusion that we have been appointed policeman to the human race.
WALTER LIPPMANN

It is our true policy to steer clear of permanent alliances with any portion of the foreign world.
GEORGE WASHINGTON

Let us never negotiate out of fear. But let us never fear to negotiate.
JOHN F. KENNEDY

Threat systems are the basis of politics as exchange systems are the basis of economics.
KENNETH BOULDING

A nation does not have to be cruel to be tough.
FRANKLIN D. ROOSEVELT

Whenever you accept our views we shall be in full agreement with you.
MOSHE DAYAN

Alliances are held together by fear, not by love.
HAROLD MACMILLAN

A nation cannot remain great if it betrays its allies and lets down its friends.
RICHARD M. NIXON

Diplomats are just as essential to starting a war as soldiers are to finishing it. You take Diplomacy out of war and the thing would fall flat in a week.
WILL ROGERS

All diplomacy is a continuation of war by other means.
CHOU EN-LAI

There are always three choices—war, surrender, and present policy.
HENRY KISSINGER

I sometimes think that strategy is nothing but tactics talked through a brass hat.
R.V. JONES

Force and Fraud are in war the two cardinal virtues.
THOMAS HOBBES

Not believing in force is the same as not believing in gravity.
LEON TROTSKY

There is nothing that war has ever achieved we could not better achieve without it.
HAVELOCK ELLIS

Speak softly and carry a big stick.
THEODORE ROOSEVELT

You can get more with a kind word and a gun than you can with a kind word alone.
JOHNNY CARSON

Diplomacy without armaments is like music without instruments.
FREDERICK THE GREAT, *of Prussia*

Let us not be deceived—we are today in the midst of a cold war.
BERNARD M. BARUCH, *1947*

Cold wars cannot be conducted by hot heads.
WALTER LIPPMANN

We are living in a pre-war and not a post-war world.
EUGENE ROSTOW

Either the Russians are doing it and therefore we must do it in order to avoid falling behind, or the Russians are not doing it and therefore we must in order to stay ahead.
PATRICIA SCHROEDER

We must recognize the chief characteristic of the modern era—a permanent state of what I call violent peace.
ADMIRAL JAMES D. WATKINS

You have a row of dominoes set up, you knock over the first one, and what will happen to the last one is . . . that it will go over very quickly.
DWIGHT D. EISENHOWER

I have never advocated war except as a means of peace.
ULYSSES S. GRANT

A wiser rule would be to make up your mind soberly what you want, peace or war, and then to get ready for what you want; for what we prepare for is what we shall get.
WILLIAM GRAHAM SUMNER

Qui desiderat pacem, praeparet bellum.
Let him who desires peace, prepare for war.
VEGETIUS

Of the four wars in my lifetime none came about because the United States was too strong.
RONALD REAGAN

Only when our arms are sufficient beyond doubt can we be certain that they will never be employed.
JOHN F. KENNEDY

I've told you before and I'll tell it to you again. The strong survive and the weak disappear. We do not propose to disappear.
JIMMY HOFFA

The longer deterrence succeeds, the more difficult it is to demonstrate what made it work.
HENRY KISSINGER

To be prepared for war is one of the most effectual means of preserving peace.
GEORGE WASHINGTON

Nations do not arm for war. They arm to keep themselves from war.
BARRY GOLDWATER

Men rise from one ambition to another; first they seek to secure themselves from attack, and then they attack others.
NICCOLO MACHIAVELLI

For a people who are free, and who mean to remain so, a well-organized and armed militia is their best security.
THOMAS JEFFERSON

The spirit of this country is totally adverse to a large military force.
THOMAS JEFFERSON

One sword keeps another in the sheath.
GEORGE HERBERT

There is no record in history of a nation that ever gained anything valuable by being unable to defend itself.

H.L. MENCKEN

Obsolete weapons do not deter.

MARGARET THATCHER

It is now quite clear that building up military power makes no country omnipotent. What is more, one-sided reliance on military power ultimately weakens other components of national security.

MIKHAIL GORBACHEV

That they may have a little peace, even the best dogs are compelled to snarl occasionally.

WILLIAM FEATHER

In this age when there can be no losers in peace and no victors in war, we must recognize the obligation to match national strength with national restraint.

LYNDON B. JOHNSON

The use or threat of force no longer can or must be an instrument of foreign policy. . . . All of us, and primarily the stronger of us, must exercise self-restraint and totally rule out any outward-oriented use of force.

MIKHAIL GORBACHEV

We cannot count on the instinct for survival to protect us against war.
RONALD REAGAN

Domestic policy can only defeat us; foreign policy can kill us.
JOHN F. KENNEDY

No nation ever had an army large enough to guarantee it against attack in time of peace, or ensure it of victory in time of war.
CALVIN COOLIDGE

As long as armies exist, any serious conflict will lead to war. A pacifism which does not actively fight against the armament of nations is and must remain impotent.
ALBERT EINSTEIN

It is the greatest possible mistake to mix up disarmament with peace. When you have peace you will have disarmament.
WINSTON CHURCHILL

Wars frequently begin ten years before the first shot is fired.
K.K.V. CASEY

How is the world ruled and how do wars start? Diplomats tell lies to journalists and then believe what they read.
KARL KRAUS

Can anything be more ridiculous than that a man has a right to kill me because he lives on the other side of the water, and because his ruler has a quarrel with mine, although I have none with him?
BLAISE PASCAL

An empire founded by war has to maintain itself by war.
MONTESQUIEU

All great civilizations, in their early stages, are based on success in war.
SIR KENNETH CLARK

War is an obscene blot on the face of the human race.
DEAN RUSK

Although war is evil, it is occasionally the lesser of two evils.
MCGEORGE BUNDY

Ambition is the grand enemy of all peace.
JOHN COWPER POWYS

War will exist until that distant day when the conscientious objector enjoys the same reputation and prestige that the warrior does today.
JOHN F. KENNEDY

All men have in them an instinct for conflict; at least all healthy men.
HILAIRE BELLOC

The tendency to aggression is an innate, independent, instinctual disposition in man . . . it constitutes the most powerful obstacle to culture.
SIGMUND FREUD

War will never cease until babies begin to come into the world with larger cerebrums and smaller adrenal glands.
H.L. MENCKEN

War is a biological necessity.
FRIEDRICH A.J. VON BERNHARDI

War is to man what maternity is to a woman. From a philosophical and doctrinal viewpoint, I do not believe in perpetual peace.
BENITO MUSSOLINI

Missiles will be able to do anything bombers can do—cheaper.
ROBERT MCNAMARA

Most quarrels are inevitable at the time; incredible afterwards.

E.M. FORSTER

Men are at war with one another because each man is at war with himself.

FRANCIS J.G. MEEHAN

The principal cause of war is war itself.

C. WRIGHT MILLS

Most sorts of diversion in men, children and other animals, are in imitation of fighting.

JONATHAN SWIFT

If the people raise a howl against my barbarity and cruelty, I will answer that war is war, and not popularity-seeking. If they want peace, they and their relatives must stop the war.

GENERAL WILLIAM SHERMAN

From fanaticism to barbarism is only one step.

DENIS DIDEROT

Just as every conviction begins as a whim so does every emancipator serve his apprenticeship as a crank. A fanatic is a great leader who is just entering the room.

HEYWOOD HALE BROUN

It is not the neutrals or the lukewarms who make history.
ADOLF HITLER

Neutrals never dominate events. They always sink. Blood alone moves the wheels of history.
BENITO MUSSOLINI

Was peace maintained by the risk of war, or because the adversary never intended aggression in the first place?
HENRY KISSINGER

We all muddled into war.
DAVID LLOYD GEORGE

The military don't start wars. The politicians start wars.
GENERAL WILLIAM WESTMORELAND

There's no difference between one's killing and making decisions that will send others to kill. It's exactly the same thing, or even worse.
GOLDA MEIR

Old men declare war. But it is the youth that must fight and die. And it is youth who must inherit the tribulation, the sorrow, and the triumphs that are the aftermath of war.
HERBERT C. HOOVER

No one man nor group of men incapable of fighting or exempt from fighting should in any way be given the power, no matter how gradually it is given them, to put this country or any country into war.

ERNEST HEMINGWAY

He had grown up in a country run by politicians who sent the pilots to man the bombers to kill the babies to make the world safer for children to grow up in.

URSULA K. LE GUIN

The world must be made safe for democracy.

WOODROW WILSON

There is only one purpose to which a whole society can be directed by a deliberate plan. That purpose is war, and there is no other.

WALTER LIPPMANN

Nothing unites the English like war. Nothing divides them like Picasso.

HUGH MILLS

Happy is that city which in time of peace thinks of war.

Inscription in the armory of Venice

The functions of a citizen and a soldier are inseparable.

BENITO MUSSOLINI

Preparation for war is a constant stimulus to suspicion and ill will.

JAMES MONROE

When the leaders speak of peace
The common folk know
That war is coming.
When the leaders curse war
The mobilization order is already written out.

BERTOLT BRECHT

If you would rule the world quietly, you must keep it amused.

RALPH WALDO EMERSON

We must remember that in time of war what is said on the enemy's side of the front is always propaganda and what is said on our side of the front is truth and righteousness, the cause of humanity and a crusade for peace.

WALTER LIPPMANN

It is a very dangerous thing to organize the patriotism of a nation if you are not sincere.

ERNEST HEMINGWAY

Propaganda is a soft weapon: hold it in your hands too long, and it will move about like a snake, and strike the other way.
JEAN ANOUILH

The first casualty when war comes is truth.
HIRAM JOHNSON

In war, truth is the first casualty.
AESCHYLUS

It is the merit of a general to impart good news, and to conceal the truth.
SOPHOCLES

Man has no nobler function than to defend the truth.
RUTH MCKENNEY

Four hostile newspapers are more to be feared than a thousand bayonets.
NAPOLEON BONAPARTE

We all had to weigh, in the balance, the difference between lives and lies.
COLONEL OLIVER NORTH

A thing is not necessarily true because a man dies for it.
OSCAR WILDE

To die for an idea is to set a rather high price on conjecture.
ANATOLE FRANCE

"Dying for an idea" again, sounds well enough, but why not let the idea die instead of you?
PERCY WYNDHAM LEWIS

In a war of ideas it is people who get killed.
STANISLAW LEC

It is easier to fight for one's principles than to live up to them.
ALFRED ADLER

Whether you are really right or not doesn't matter, it's the belief that counts.
ROBERTSON DAVIES

The wrong war, at the wrong place, at the wrong time, and with the wrong enemy.
GENERAL OMAR BRADLEY, *on the proposal to carry the Korean conflict into China*

In war, moral considerations make up three-quarters of the game: the relative balance of manpower accounts only for the remaining quarter.
NAPOLEON BONAPARTE

A man may devote himself to death and destruction to save a nation; but no nation will devote itself to death and destruction to save mankind.
SAMUEL TAYLOR COLERIDGE

Man is the only animal of which I am thoroughly and cravenly afraid. . . . There is no harm in a well-fed lion. It has no ideals, no sect, no party, no nation, no class: in short no reason for destroying anything it does not want to eat.
GEORGE BERNARD SHAW

Beasts do not fight collectively. Who has ever seen ten lions fight ten bulls? Yet how often do 20,000 armed Christians fight 20,000 armed Christians?
ERASMUS

It is a perplexing and pleasant truth that when men already have something worth fighting for, they do not feel like fighting.
ERIC HOFFER

To fight for a reason and in a calculating spirit is something your true warrior despises.
GEORGE SANTAYANA

If man is not ready to risk his life, where is his dignity?
ANDRÉ MALRAUX

Better that we should die fighting than be outraged and dishonored. Better to die than to live in slavery.
EMMELINE PANKHURST

It is a fearful thing to lead this great peaceful people into war, into the most terrible and disastrous of all wars, civilization itself seeming to be in the balance. But the right is more precious than peace, and we shall fight for the things which we have always carried nearest our hearts—for democracy.
WOODROW WILSON

The only war I ever approved of was the Trojan War; it was fought over a woman and the men knew what they were fighting for.
WILLIAM LYON PHELPS

The world is a fine place and worth fighting for.
ERNEST HEMINGWAY

I only regret that I have but one life to lose for my country.
NATHAN HALE

If a country is worth living in, it is worth fighting for.
MANNING COLES

Our country! In her intercourse with foreign nations, may she always be in the right; but our country, right or wrong.
STEPHEN DECATUR

"My country right or wrong" is like saying, "My mother drunk or sober."
G.K. CHESTERTON

The most persistent sound which reverberates through men's history is the beating of the war drum.
ARTHUR KOESTLER

As a rule, high culture and military power go hand in hand, as evidenced in the cases of Greece and Rome.
BARON COLMAR VON DER GOLTZ

Human war has been the most successful of our cultural traditions.
ROBERT ARDREY

War should belong to the tragic past, to history: it should find no place on humanity's agenda for the future.
POPE JOHN PAUL II

History is littered with wars which everybody knew would never happen.
ENOCH POWELL

Fiddle-dee-dee. War, war, war. This war talk's spoiling all the fun at every party this spring. I get so bored I could scream. Besides, there isn't going to be any war!

VIVIEN LEIGH, *as Scarlett O'Hara in* Gone With the Wind

History teaches that wars begin when governments believe the price of aggression is cheap.

RONALD REAGAN

The whole history of the world is summed up in the fact that, when nations are strong, they are not always just, and when they wish to be just, they are no longer strong.

WINSTON CHURCHILL

A government needs one hundred soldiers for every guerilla it faces.

FULGENCIO BATISTA

Political power grows out of the barrel of a gun.

MAO TSE-TUNG

The guerilla fights the war of the flea, and his military enemy suffers the dog's disadvantages: too much to defend; too small, ubiquitous and agile an enemy to come to grips with.

ROBERT TABER

The conventional army loses if it does not win. The guerilla wins if he does not lose.

HENRY KISSINGER

Those who make peaceful revolution impossible will make violent revolution inevitable.

JOHN F. KENNEDY

A people that wants to win its independence cannot limit itself to ordinary means of war. Uprisings in mass, revolutionary war, guerillas everywhere, that is the only means through which a small nation can get the better of a big one, a less strong army be put in a position to resist a stronger and better organized one.

FRIEDRICH ENGELS

A foreign war is a scratch on the arm; a civil war is an ulcer which devours the vitals of a nation.

VICTOR HUGO

Revolutions are not made by fate but by men.

JACOB BRONOWSKI

The Civil War is not ended: I question whether any serious civil war ever does end.

T.S. ELIOT

A revolution is interesting insofar as it avoids like the plague the plague it promised to heal.
DANIEL BERRIGAN

Revolution is like one cocktail; it just gets you organized for the next.
WILL ROGERS

At least we're getting the kind of experience we need for the next war.
ALLEN DULLES

A revolution is not a bed of roses. A revolution is a struggle to the death between the future and the past.
FIDEL CASTRO

Few revolutionists would be such if they were heirs to a baronetcy.
GEORGE SANTAYANA

Revolutions have never lightened the burden of tyranny, they have only shifted it to another shoulder.
GEORGE BERNARD SHAW

Every revolutionary ends up by becoming either an oppressor or a heretic.
ALBERT CAMUS

There is hardly such a thing as a war in which it makes no difference who wins. Nearly always one side stands more or less for progress, the other side more or less for reaction.
GEORGE ORWELL

It was duty, honor, country . . . our country had been attacked. . . . It was freedom vs. oppression. It was against imperialism and against fascism, and the country was so together, and I wanted to be on the cutting edge.
GEORGE BUSH, *on his decision to enlist in 1941*

What difference does it make to the dead, the orphans, and the homeless, whether the mad destruction is wrought under the name of totalitarianism or the holy name of liberty or democracy?
MAHATMA GANDHI

What the hell difference does it make, left or right? There were good men lost on both sides.
BRENDAN BEHAN

As long as war is regarded as wicked, it will always have its fascination. When it is looked upon as vulgar, it will cease to be popular.
OSCAR WILDE

All wars are popular for the first thirty days.
ARTHUR SCHLESINGER, JR.

The troops will march in, the bands will play, the crowds will cheer, and in four days everyone will have forgotten. Then we will be told we have to send in more troops. It's like taking a drink. The effect wears off, and you have to take another.
JOHN F. KENNEDY

War is, after all, the universal perversion.
JOHN RAE

War alone brings up to its highest tension all human energy, and puts the stamp of nobility upon the peoples who have the courage to meet it. All other trials are substitutes, which never really put men into the position where they have to make the great decision—the alternatives of life or death.
BENITO MUSSOLINI

War is like love, it always finds a way.
BERTOLT BRECHT

War hath no fury like a non-combatant.
C.E. MONTAGUE

Usually, when a lot of men get together, it's called war.
MEL BROOKS

War is nothing but a duel on a larger scale.
KARL VON CLAUSEWITZ

To those for whom war is necessary, it is just; and resort to arms is righteous for those to whom no further hope remains.
LIVY

Lucky are soldiers who strive in a just war; for them it is an easy entry into heaven.
Bhagavad Gita

Moderation in war is imbecility.
ADMIRAL JOHN FISHER

He knew that the essence of war is violence, and that moderation in war is imbecility.
THOMAS BABINGTON MACAULAY

Every attempt to make war easy and safe will result in humiliation and disaster.
GENERAL WILLIAM SHERMAN

I sincerely wish war was a pleasanter and easier business than it is, but it does not admit of holidays.
ABRAHAM LINCOLN, *on the suggestion that he take a holiday*

War is the greatest plague that can afflict humanity; it destroys religion, it destroys states, it destroys families. Any scourge is preferable to it.
MARTIN LUTHER

How vile and despicable war seems to me! I would rather be hacked in pieces than take part in such an abominable business.
ALBERT EINSTEIN

Have you ever thought that war is a madhouse and that everyone in the war is a patient?
ORIANA FALLACI

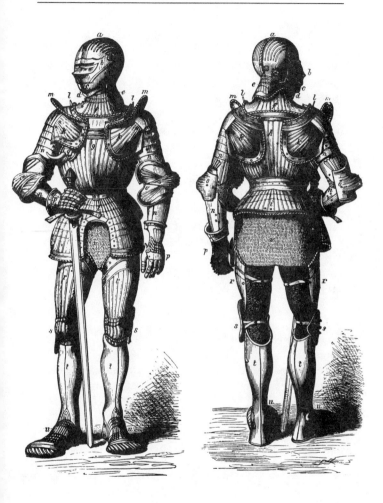

In war more than anywhere else in the world things happen differently from what we had expected, and look differently when near from what they did at a distance.
KARL VON CLAUSEWITZ

In war trivial causes produce momentous events.
JULIUS CAESAR

War is the unfolding of miscalculations.
BARBARA TUCHMAN

War is mainly a catalogue of blunders.
WINSTON CHURCHILL

War is a series of catastrophes which result in victory.
GEORGES CLEMENCEAU

To inquire if and where we made mistakes is not to apologize. War is replete with mistakes because it is full of improvisations. In war we are always doing something for the first time. It would be a miracle if what we improvised under the stress of war should be perfect.
VICE ADMIRAL HYMAN RICKOVER

I detest war. It spoils armies.
GRAND DUKE CONSTANTINE OF RUSSIA

War involves in its progress such a train of unforeseen and unsupposed circumstances that no human wisdom can calculate the end. It has but one thing certain, and that is to increase taxes.

THOMAS PAINE

War makes strange giant creatures out of us little routine men who inhabit the earth.

ERNIE PYLE

We are not at war with Egypt. We are in a state of armed conflict.

ANTHONY EDEN

It is far easier to make war than to make peace.

GEORGES CLEMENCEAU

The object of war is not to die for your country but to make the other bastard die for his.
GENERAL GEORGE PATTON

War is war. The only good human being is a dead one.
GEORGE ORWELL

The power to wage war is the power to wage war successfully.
CHARLES EVANS HUGHES

One knows what a war is about only when it is over.
H.N. BRAILSFORD

War is, at first, the hope that one will be better off; next, the expectation that the other fellow will be worse off; then, the satisfaction that he isn't any better off; and, finally, the surprise at everyone's being worse off.
KARL KRAUS

War means ugly mob-madness, crucifying the truth-tellers, choking the artists, sidetracking reforms, revolutions, and the working of social forces.
JOHN REED

It simply is not true that war never settles anything.
FELIX FRANKFURTER

Hate is able to provoke disorders, to ruin a social organization, to cast a country into a period of bloody revolutions; but it produces nothing.
GEORGES SOREL

It is well that war is so terrible—we would grow too fond of it.
ROBERT E. LEE

We are all familiar with the argument: Make war dreadful enough, and there will be no war. And we none of us believe it.
JOHN GALSWORTHY

My dynamite will sooner lead to peace
Than a thousand world conventions.
As soon as men will find that in one instant
Whole armies can be utterly destroyed,
They surely will abide by golden peace.
ALFRED BERNHARD NOBEL

There is many a boy here today who looks on war as all glory, but boys, it is all hell.
GENERAL WILLIAM SHERMAN

War is not an adventure. It is a disease. It is like typhus.
ANTOINE DE SAINT-EXUPÉRY

War would end if the dead could return.
STANLEY BALDWIN

They were going to look at war, the red animal—
war, the blood-swollen god.
STEPHEN CRANE

I have seen war. I have seen war on land and sea. I
have seen blood running from the wounded. I have
seen men coughing out their gassed lungs. I have
seen the dead in the mud. I have seen cities destroyed. . . . I have seen children starving. I have
seen the agony of mothers and wives. I hate war.
FRANKLIN D. ROOSEVELT

Where are your legs that used to run,
When you went for to carry a gun,
Indeed your dancin' days are done,
Johnny I hardly knew ye.
"Johnny I Hardly Knew Ye," Irish folk song

We have resolved to endure the unendurable and
suffer what is insufferable.
HIROHITO, *after the Hiroshima bombing*

I hate war as only a soldier who has lived it can,
only as one who has seen its brutality, its futility, its
stupidity.
DWIGHT D. EISENHOWER

I think I'd be a better president because I was in combat.

GEORGE BUSH

I'm inclined to think that a military background wouldn't hurt anyone.

WILLIAM FAULKNER

Most of the miseries of the world were caused by wars. And, when the wars were over, no one ever knew what they were about.

LESLIE HOWARD, *as Ashley Wilkes in* Gone With the Wind

But that was war. Just about all he could find in its favor was that it paid well and liberated children from the pernicious influence of their parents.

JOSEPH HELLER

The only defense is offense, which means that you have to kill more women and children more quickly than the enemy if you wish to save yourselves.

STANLEY BALDWIN

Everyone is always talking about our defense effort in terms of defending women and children, but no one ever asks the women and children what they think.

PATRICIA SCHROEDER

Let me have war, say I; it exceeds peace as far as day does night; it's spritely, waking, audible, and full of vent. Peace is a very apoplexy, lethargy: mulled, deaf, sleepy, insensible; a getter of more bastard children than war's a destroyer of men.

    WILLIAM SHAKESPEARE, Coriolanus

War is delightful to those who have had no experience of it.

    ERASMUS

The grass grows green on the battlefield, but never on the scaffold.

    WINSTON CHURCHILL

Delete any footage which includes the idea that war is not altogether glamorous and noble.

    JOSEPH I. BREEN, *Hollywood film association executive*

War may make a fool of man, but it by no means degrades him; on the contrary, it tends to exalt him, and its net effects are much like those of motherhood on women.

    H.L. MENCKEN

I heard the bullets whistle, and believe me, there is something charming in the sound.

    GEORGE WASHINGTON

I love the smell of napalm in the morning. . . . It smells like victory.

ROBERT DUVALL, *as Lieutenant Colonel Kilgore in* Apocalypse Now

It is not merely cruelty that leads men to love war, it is excitement.

HENRY WARD BEECHER

The statistics of suicide show that, for non-combatants at least, life is more interesting in war than in peace.

WILLIAM RALPH INGE

The only people who ever loved war for long were profiteers, generals, staff officers and whores.
ERNEST HEMINGWAY

I love war and responsibility and excitement. Peace is going to be hell on me.
GENERAL GEORGE PATTON

War, when you are at it, is horrible and dull. It is only when time has passed that you see that its message was divine.
OLIVER WENDELL HOLMES, JR.

'Tis better to have fought and lost,
Than never to have fought at all.
ARTHUR HUGH CLOUGH

People who are vigorous and brutal often find war enjoyable, provided that it is a victorious war and that there is not too much interference with rape and plunder. This is a great help in persuading people that wars are righteous.
BERTRAND RUSSELL

I think it is reasonable that if we must continue to fight wars, they ought to be fought by those people who really want to fight them. Since it seems to be the top half of the generation gap that is the most enthusiastic about going to war, why not send the Old Folks Brigade to Vietnam—with John Wayne leading them?
DICK GREGORY

I know I am among civilized men because they are fighting so savagely.
VOLTAIRE

No amount of study or learning will make a man a leader unless he has the natural qualities of one.
SIR ARCHIBALD WAVELL

He who wishes to be obeyed must know how to command.
NICCOLO MACHIAVELLI

Learn to obey before you learn to command.
SIR IAN HAMILTON

I can no longer obey; I have tasted command, and I cannot give it up.
NAPOLEON BONAPARTE

To command is to serve, nothing more, nothing less.
ANDRÉ MALRAUX

A chief is a man who assumes responsibility. He says, "I was beaten," he does not say "My men were beaten."
ANTOINE DE SAINT-EXUPÉRY

To be a leader of men one must turn one's back on men.
HAVELOCK ELLIS

I must follow them. I am their leader.
ANDREW BONAR LAW

The most important quality in a leader is that of being acknowledged as such. All leaders whose fitness is questioned are clearly lacking in force.
ANDRÉ MAUROIS

If, in order to succeed in an enterprise, I were obliged to choose between fifty deer commanded by a lion, and fifty lions commanded by a deer, I should consider myself more certain of success with the first group than with the second.
SAINT VINCENT DE PAUL

Everyone imposes his own system as far as his army can reach.
JOSEPH STALIN

There are no warlike people—just warlike leaders.
RALPH BUNCHE

The leader of genius must have the ability to make different opponents appear as if they belonged to one category.

ADOLF HITLER

To a surprising extent the war-lords in shining armour, the apostles of martial virtues, tend not to die fighting when the time comes. History is full of ignominious getaways by the great and famous.

GEORGE ORWELL

I tell you Wellington is a bad general, the English are bad soldiers; we will settle the matter by lunch time.

NAPOLEON BONAPARTE, *the morning of the Battle of Waterloo*

It is not the business of generals to shoot one another.

ARTHUR WELLESLEY, DUKE OF
WELLINGTON

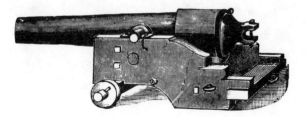

Name me an emperor who was ever struck by a cannonball.

CHARLES V, *of France*

War is too important to be left to the generals.

GEORGES CLEMENCEAU

Nothing remains static in war or in military weapons, and it is consequently often dangerous to rely on courses suggested by apparent similarities in the past.

ADMIRAL ERNEST JOSEPH KING

Nothing so comforts the military mind as the maxim of a great, but dead, general.

BARBARA TUCHMAN

The only thing harder than getting a new idea into the military mind is to get an old one out.

B.H. LIDDELL HART

The greatest general is he who makes the fewest mistakes.

NAPOLEON BONAPARTE

I didn't fire him [General MacArthur] because he was a dumb son of a bitch, although he was, but that's not against the law for generals. If it was, half to three-quarters of them would be in jail.

HARRY TRUMAN

Ah! The Generals! They are numerous but not good for much!

ARISTOPHANES

Generals think war should be waged like the tourneys of the Middle Ages. I have no use for knights; I need revolutionaries.

ADOLF HITLER

As for being a General, well, at the age of four with paper hats and wooden swords we're all Generals. Only some of us never grow out of it.

PETER USTINOV

If I had learned to type, I never would have made brigadier general.

BRIGADIER GENERAL ELIZABETH P.
HOISINGTON

I am convinced that the best service a retired general can perform is to turn in his tongue along with his suit and to mothball his opinion.

GENERAL OMAR BRADLEY

Britain is no longer a world power—all they have left are generals and admirals and bands.

GENERAL GEORGE BROWN

He smiles with the faraway, sea-remembering smile of all desk admirals.
PETER GRIER

An admiral has to be put to death now and then to encourage the others.
VOLTAIRE

It is dangerous to meddle with Admirals when they say they can't do things. They have always got the weather or fuel or something to argue about.
WINSTON CHURCHILL

In my experience . . . officers with high athletic qualifications are not usually successful in the higher ranks.
WINSTON CHURCHILL

Every officer has his ceiling in rank, beyond which he should not be allowed to rise—particularly in war-time.
FIELD MARSHAL BERNARD MONTGOMERY

An army cannot be administered. It must be led.
FRANZ JOSEPH STRAUSS

Paper-work will ruin any military force.
LIEUTENANT GENERAL LEWIS "CHESTY" PULLER

The little affair of operational command is something anybody can do.
ADOLF HITLER

A bulky staff implies a division of responsibility, slowness of action, and indecision, whereas a small staff implies activity and concentration of purpose.
GENERAL WILLIAM SHERMAN

Surround yourself with the best people you can find, delegate authority, and don't interfere.
RONALD REAGAN

All very successful commanders are prima donnas and must be so treated.
GENERAL GEORGE PATTON

In military affairs, only military men should be listened to.
THEODORE ROOSEVELT

An important difference between the military operation and a surgical operation is that the patient is not tied down. But it is a common fault of generalship to assume that he is.

B.H. LIDDELL HART

The first advance I am going to give my successor is to watch the generals and to avoid feeling that just because they were military men their opinions on military matters were worth a damn.

JOHN F. KENNEDY

Soldiers win battles and generals get the credit.

NAPOLEON BONAPARTE

Sir, my strategy is one against ten, my tactics ten against one.

ARTHUR WELLESLEY, DUKE OF
WELLINGTON

I can always make it a rule to get there first with the most men.

NATHAN BEDFORD FORREST

It is no use to get there first unless, when the enemy arrives, you have also the greater men—the greater force.

ADMIRAL ALFRED THAYER MAHAN

If they've been put there to fight, there are far too few. If they've been put there to be killed, there are far too many.

ERNEST F. HOLLINGS

The first law of war is to preserve ourselves and destroy the enemy.

MAO TSE-TUNG

The U.S. has broken the second rule of war. That is, don't go fighting with your land army on the mainland of Asia. Rule One is don't march on Moscow. I developed these two rules myself.

FIELD MARSHAL BERNARD MONTGOMERY

It is fatal to enter any war without the will to win it.

GENERAL DOUGLAS MACARTHUR

If you start to take Vienna—take Vienna.

NAPOLEON BONAPARTE

Whoever conquers a free town and does not demolish it commits a great error and may expect to be ruined himself.

NICCOLO MACHIAVELLI

You don't hurt 'em if you don't hit 'em.

LIEUTENANT GENERAL LEWIS "CHESTY" PULLER

Do not hit at all if it can be avoided, but never hit softly.
THEODORE ROOSEVELT

There are not fifty ways of fighting, there is only one way: to be the conqueror.
ANDRÉ MALRAUX

Power is not revealed by striking hard or often, but by striking true.
HONORÉ DE BALZAC

Treating your adversary with respect is striking soft in battle.
SAMUEL JOHNSON

There is such a thing as legitimate warfare: war has its laws; there are things which may fairly be done, and things which may not be done.
JOHN HENRY CARDINAL NEWMAN

To lead an untrained people to war is to throw them away.
CONFUCIUS

Every citizen [should] be a soldier. This was the case with the Greeks and the Romans, and must be that of every free state.
THOMAS JEFFERSON

The general of a large army may be defeated, but you cannot defeat the determined mind of a peasant.

CONFUCIUS

The army taught me some great lessons—to be prepared for catastrophe—to endure being bored—and to know that however fine a fellow I thought myself in my usual routine there were other situations in which I was inferior to men that I might have looked down upon had not experience taught me to look up.

OLIVER WENDELL HOLMES, JR.

Being in the army is like being in the Boy Scouts, except that the Boy Scouts have adult supervision.

BLAKE CLARK

And here is the lesson I learned in the army. If you want to do a thing badly, you have to work at it as though you want to do it well.

PETER USTINOV

My dear McClellan:
If you don't want to use the Army I should like to borrow it for a while.

ABRAHAM LINCOLN

An army marches on its stomach.

NAPOLEON BONAPARTE

Join the army, see the world, meet interesting people, and kill them.

Unknown

We joined the Navy to see the world
And what did we see?
We saw the sea.

IRVING BERLIN

The Army is always the same. The sun and the moon change, but the Army knows no seasons.

JOHN WAYNE, *as Captain Nathan Brittles in* She Wore a Yellow Ribbon

It is significant that despite claims of air enthusiasts no battleship has yet been sunk by bombs.

Caption to a photograph of the U.S.S. Arizona *in the Army–Navy football game program, November 29, 1941*

The Navy's a very gentlemanly business. You fire at the horizon to sink a ship and then you pull people out of the water and say, "Frightfully sorry, old chap."

WILLIAM GOLDING

It is highly unlikely that an airplane, or a fleet of them, could ever sink a fleet of Navy vessels under battle conditions.

FRANKLIN D. ROOSEVELT

Don't talk to me about naval tradition. It's nothing but rum, sodomy, and the lash.

WINSTON CHURCHILL

Something must be left to chance; nothing is sure in a sea fight beyond all others.

HORATIO NELSON

No man will be a sailor who has contrivance enough to get himself into a jail; for being in a ship is being in a jail, with the chance of being drowned. A man in jail has more room, better food, and commonly better company.

SAMUEL JOHNSON

In an English ship, they say, it is poor grub, poor pay, and easy work; in an American ship, good grub, good pay, and hard work. And this is applicable to the working populations of both countries.

JACK LONDON

The Navy is a machine invented by geniuses, to be run by idiots.

HERMAN WOUK, The Caine Mutiny

Why should we have a navy at all? There are no enemies for it to fight except apparently the Army Air Force.

GENERAL CARL SPAATZ

Cease firing, but if any enemy planes appear, shoot them down in a friendly fashion.

ADMIRAL WILLIAM "BULL" HALSEY, JR.

I have always regarded the forward edge of the battlefield as the most exclusive club in the world.

SIR BRIAN HORROCKS

I hated the bangs in the war: I always felt a silent war should have been far more tolerable.
PAMELA HANSFORD JOHNSON

Music played at weddings always reminds me of the music played for soldiers before they go into battle.
HEINRICH HEINE

That a man can take pleasure in marching in fours to the strains of a band is enough to make me despise him.
ALBERT EINSTEIN

How good bad music and bad reasons sound when we march against an enemy.
FRIEDRICH NIETZSCHE

Military justice is to justice what military music is to music.
GEORGES CLEMENCEAU

Stand your ground. Don't fire unless fired upon, but if they mean to have a war, let it begin here!
JOHN PARKER

Don't fire until you see the whites of their eyes.
WILLIAM PRESCOTT

Elevate those guns a little lower.
ANDREW JACKSON

. . . the race *is* not to the swift, nor the battle to the strong . . .
Ecclesiastes 9:11

Peerless, fearless, an army's flower!
Sterner soldiers the world never saw,
Marching lightly, that summer's hour,
To death and failure and fame forever.
HELEN GRAY CONE

We all need someone we can bleed on.
MICK JAGGER

In 40 hours I shall be in battle, with little information, and on the spur of the moment will have to make most momentous decisions. But I believe that one's spirit enlarges with responsibility and that, with God's help, I shall make them and make them right.
GENERAL GEORGE PATTON

After a battle is over people talk a lot about how decisions were methodically reached, but actually there's always a hell of a lot of groping around.
ADMIRAL FRANK FLETCHER

The core of the military profession is discipline and the essence of discipline is obedience. Reasonable orders are easy enough to obey; it is capricious, bureaucratic or plain idiotic demands that form the habit of discipline.

BARBARA TUCHMAN

In action it is better to order than to ask.

SIR IAN HAMILTON

A good soldier has his heart and soul in it. When he receives an order, he gets a hard-on, and when he sends his lance into the enemy's guts, he comes.

BERTOLT BRECHT

Wars may be fought with weapons, but they are won by men. It is the spirit of the men who follow and of the man who leads that gains the victory.

GENERAL GEORGE PATTON

Men do not take good iron to make nails nor good men to make soldiers.

PEARL S. BUCK

You can always tell an old soldier by the insides of his holsters and cartridge boxes. The young ones carry pistols and cartridges: the old ones, grub.

GEORGE BERNARD SHAW

Old soldiers never die;
  They only fade away!
    "War Song of the British Soldiers"

Old soldiers never die; they just fade away.
    GENERAL DOUGLAS MACARTHUR, *in a 1951*
    address to a joint session of Congress

When you put on a uniform, there are certain inhibitions that you accept.
    DWIGHT D. EISENHOWER

A good uniform must work its way with the women, sooner or later.
    CHARLES DICKENS

The [military] officer is a being apart, a kind of artist breathing the grand air in the brilliant profession of arms, in a uniform that is always seductive. . . . To me the officer is a separate race.

MATA HARI

Three-quarters of a soldier's life is spent in aimlessly waiting about.

EUGEN ROSENSTOCK-HUESSY

The chief attraction of military service has consisted and will consist in this compulsory and irreproachable idleness.

LEO TOLSTOY

We played remote bases, the kind of bases where guys went to bed with their rifles by their sides; not for safety, but for companionship.

BOB HOPE

I never expect a soldier to think.

GEORGE BERNARD SHAW

Military intelligence is a contradiction in terms.

GROUCHO MARX

Man uses his intelligence less in the care of his own species than he does in the care of anything else he governs.

ABRAHAM MEYERSON

The services in wartime are fit only for desperadoes, but in peace are fit only for fools.
 BENJAMIN DISRAELI

In order to have good soldiers a nation must always be at war.
 NAPOLEON BONAPARTE

The British soldier can stand up to anything except the British War Office.
 GEORGE BERNARD SHAW

The peril of the hour moved the British to tremendous exertions, just as always in a moment of extreme danger things can be done which had previously been thought impossible. Mortal danger is an effective antidote for fixed ideas.
 GENERAL ERWIN ROMMEL

The sense of danger is never, perhaps, so fully apprehended as when the danger has been overcome.
 SIR ARTHUR HELPS

Theirs is not to reason why,
Theirs is but to do or die.
 ALFRED, LORD TENNYSON, *"The Charge of the Light Brigade"*

A bayonet is a weapon with a worker at both ends.
 British pacifist slogan

I would lay down my life for America, but I cannot trifle with my honor.

JOHN PAUL JONES

It is better to die on your feet than to live on your knees.

EMILIANO ZAPATA

The officers will take all proper opportunities to inculcate in their men's minds a reliance on the bayonet, men of their bodily strength and even a coward may be their match in firing. But a bayonet in the hands of the valiant is irresistable.

MAJOR GENERAL JOHN BURGOYNE

Patriotism has served, at different times, as widely different ends as a razor, which ought to be used in keeping your face clean and yet may be used to cut your own throat or that of an innocent person.

C.E. MONTAGUE

"Why me?" That is the soldier's first question, asked each morning as the patrols go out and each evening as the night settles around the foxholes.

WILLIAM BROYLES, JR.

It is enough for the world to know that I am a soldier.

GENERAL WILLIAM SHERMAN

Everyone has the feeling that characterizes war. It wasn't me, see? It wasn't me.
ERNEST HEMINGWAY

The aim of military training is not just to prepare men for battle, but to make them long for it.
LOUIS SIMPSON

If it's natural to kill why do men have to go into training to learn how?

JOAN BAEZ

Putting aside all the fancy words and academic doubletalk, the basic reason for having a military is to do two jobs—to kill people and to destroy the works of man.

GENERAL THOMAS S. POWER

A prisoner of war is a man who tries to kill you and fails, and then asks you not to kill him.

WINSTON CHURCHILL

In combat, life is short, nasty and brutish. The issues of national policy which brought him into war are irrelevant to the combat soldier; he is concerned with his literal life chances.

CHARLES E. MOSKOS, JR.

They were learning the reality of war, these youngsters, getting face to face with the sickening realization that men get killed uselessly because their generals are stupid, so that desperate encounters where the last drop of courage has been given serve the country not at all and make a patriot look a fool.

BRUCE CATTON

They wish to hell they were someplace else, and they wish to hell they would get relief. They wish to hell the mud was dry and they wish to hell their coffee was hot. They want to go home. But they stay in their wet holes and fight, and then they climb out and crawl through minefields and fight some more.
BILL MAULDIN

They were called grunts, and many of them, however grudgingly, were proud of the name. They were the infantrymen, the foot soldiers of the war.
BERNARD EDELMAN

Men acquainted with the battlefield will not be found among the numbers that glibly talk of another war.
DWIGHT D. EISENHOWER

Look at an infantryman's eyes and you can tell how much war he has seen.
BILL MAULDIN

When a soldier sees a clean face there's one more whore in the world.
BERTOLT BRECHT

The most terrible job in warfare is to be a second lieutenant leading a platoon when you are on the battlefield.
DWIGHT D. EISENHOWER

Men and women who would shrink from doing anything dishonorable in the sphere of personal relationships are ready to lie and swindle and to steal and even murder when they are representing their country.

ALDOUS HUXLEY

They were all enemy. They were all to be destroyed.

LIEUTENANT WILLIAM CALLEY, JR.

Battle is the most magnificent competition in which a human being can indulge. It brings out all that is best; it removes all that is base. All men are afraid in battle. The coward is the one who lets his fear overcome his sense of duty. Duty is the essence of manhood.

GENERAL GEORGE PATTON

To delight in war is a merit in the soldier, a dangerous quality in the captain and a positive crime in the statesman.

GEORGE SANTAYANA

It's one of the most serious things that can possibly happen to one in a battle—to get one's head cut off.

LEWIS CARROLL, Through the Looking Glass

Nothing is more exhilarating than to be shot at without result.

WINSTON CHURCHILL

A revealing light is thrown on this subject through the studies by Medical Corps psychiatrists of the combat fatigue cases in the European Theater. They found that fear of killing, rather than fear of being killed, was the most common cause of battle failure, and that fear of failure ran a strong second.

S.L.A. MARSHALL

I joined the army, and succeeded in killing about as many of the enemy as they of me.

CHARLES HENRY SMITH (BILL ARP)

To choose one's victim, to prepare one's plan minutely, to slake an implacable vengeance, and then go to bed. . . . There is nothing sweeter in the world.

JOSEPH STALIN

The greatest pleasure is to vanquish your enemies and chase them before you, to rob them of their wealth and see those dear to them bathed in tears, to ride their horses and clasp to your bosom their wives and daughters.

GENGHIS KHAN

Join a Highland regiment, me boy. The kilt is an unrivaled garment for fornication and diarrhea.

LIEUTENANT COLONEL JOHN MASTERS

I can still hear the awful lamentation of the women and the drunken uproar of the men during the first days of war.

GENERAL PETRO GRIGORENKO

When the military man approaches, the world locks up its spoons and packs off its womankind.

GEORGE BERNARD SHAW

There's no such thing as a crowded battlefield. Battlefields are lonely places.

GENERAL ALFRED M. GRAY

This man Wellington is so stupid he does not know when he is beaten and goes on fighting.

NAPOLEON BONAPARTE

They've got us surrounded again, the poor bastards.

GENERAL CREIGHTON W. ABRAMS

The good company has no place for the officer who would rather be right than loved, for the time will quickly come when he walks alone, and in battle no man may succeed in solitude.

S.L.A. MARSHALL

After about 25 medals, you run out of shoulder to put them on.

COLONEL GEORGE DAY

A chest full of medals is nothing more than a resume in 3-D and Technicolor.
OWEN EDWARDS

The world continues to offer glittering prizes to those who have stout hearts and sharp swords.
FREDERICK EDWIN SMITH, *Earl of Birkenhead*

When I was in the military they gave me a medal for killing two men, and a discharge for loving one.
Inscription on the tombstone of Air Force sergeant Leonard
Matlovich who died of AIDS

Men love war because it allows them to look serious, because it's the only thing that stops women from laughing at them.
JOHN FOWLES

Ask any soldier. To kill a man is to merit a woman.
JEAN GIRAUDOUX

The last thing a woman will consent to discover in a man whom she loves or on whom she simply depends, is want of courage.
JOSEPH CONRAD

It's not what men fight for. They fight in the last resort to impress their mothers.
GABRIEL FIELDING

None but the Brave deserves the Fair.
JOHN DRYDEN

Never in the field of human conflict was so much owed by so many to so few.
WINSTON CHURCHILL, *on British airmen in the Battle of Britain*

In war the heroes always outnumber the soldiers ten to one.
H.L. MENCKEN

When it comes to the pinch, human beings are heroic.
GEORGE ORWELL

Unhappy the land that is in need of heroes.
BERTOLT BRECHT

A hero is no braver than an ordinary man, but he is brave five minutes longer.
RALPH WALDO EMERSON

The ordinary man is involved in action, the hero acts. An immense difference.
HENRY MILLER

I'm no hero. Heroes are for the late show.
SERGEANT PHILLIP ARTEBURY

This thing of being a hero, about the main thing to it is to know when to die.
WILL ROGERS

No hero is mortal till he dies.
W.H. AUDEN

It's not that I'm afraid to die. I just don't want to be there when it happens.
WOODY ALLEN

Being a hero is the shortest-lived profession on earth.
WILL ROGERS

Show me a hero and I'll write you a tragedy.
F. SCOTT FITZGERALD

And each man stands with his face in the light
  Of his own drawn sword,
Ready to do what a hero can.
    ELIZABETH BARRETT BROWNING

Courage is sustained . . . by calling up anew the vision of the goal.
    A.G. SERTILLANGES

But courage which goes against military expediency is stupidity, or, if it is insisted upon by a commander, irresponsibility.
    GENERAL ERWIN ROMMEL

Courage without conscience is a wild beast.
    ROBERT G. INGERSOLL

The first virtue in a soldier is endurance of fatigue; courage is only the second virtue.
    NAPOLEON BONAPARTE

War is fear cloaked in courage.
    GENERAL WILLIAM WESTMORELAND

Courage is resistance to fear, mastery of fear, not absence of fear.
MARK TWAIN

Courage is the fear of being thought a coward.
HORACE SMITH

Courage is fear holding on a minute longer.
GENERAL GEORGE PATTON

It sometimes helps if you sort out in your mind the very real difference between being *brave* and being *fearless*. Being brave means doing or facing something frightening. . . . Being fearless means being without fear.
PENELOPE LEACH

Bravery is the capacity to perform properly even when scared half to death.
GENERAL OMAR BRADLEY

Valor lies just halfway between rashness and cowardice.
MIGUEL DE CERVANTES

The paradox of courage is that a man must be a little careless of his life even in order to keep it.
G.K. CHESTERTON

Courage is the price that Life exacts for granting peace.

AMELIA EARHART

Fighting is like champagne. It goes to the heads of cowards as quickly as of heroes. Any fool can be brave on a battlefield when it's be brave or else be killed.

MARGARET MITCHELL

To call war the soil of courage and virtue is like calling debauchery the soil of love.

GEORGE SANTAYANA

A man who is not afraid is not aggressive, a man who has no sense of fear of any kind is really a free, a peaceful man.

JIDDU KRISHNAMURTI

One man with courage makes a majority.

ANDREW JACKSON

The courage we desire and prize is not the courage to die decently, but to live manfully.

THOMAS CARLYLE

Only the brave know how to forgive. . . . A coward never forgave; it is not in his nature.

LAURENCE STERNE

It is thus that mutual cowardice keeps us in peace. Were one half of mankind brave and one cowards, the brave would be always beating the cowards. Were all brave, they would lead a very uneasy life; all would be continually fighting; but being all cowards, we go on very well.
SAMUEL JOHNSON

In time of war the loudest patriots are the greatest profiteers.
AUGUST BEBEL

To wage war, you need first of all money; second, you need money, and third, you also need money.
PRINCE MONTECUCCOLI

The sinews of war are five—men, money, material, maintenance (food) and morale.
BERNARD M. BARUCH

The sinews of war are infinite money.
CICERO

Money is the sinews of love, as of war.
GEORGE FARQUHAR

Wars are not paid for in wartime, the bill comes later.
BENJAMIN FRANKLIN

Frankly, I'd like to see the government get out of war altogether and leave the whole field to private industry.

JOSEPH HELLER

We must guard against the acquisition of unwarranted influence . . . by the military industrial complex. The potential for the disastrous rise of misplaced power exists and will persist.

DWIGHT D. EISENHOWER

You have to remember, we don't have the military industrial complex that we once had, when President Eisenhower spoke about it.

RONALD REAGAN

War is a necessary part of God's arrangement of the world. . . . Without war, the world would slide dissolutely into materialism.

HELMUTH VON MOLTKE

I tell you how it should all be done. Whenever there's a big war coming, you should rope off a big field and sell tickets. And, on the big day, you should take all the kings and cabinets and their generals, put them in the center dressed in their underpants and let them fight it out with clubs.

LOUIS WOLHEIM, *as Katczinsky in* All Quiet on the Western Front

When the rich wage war, it is the poor who die.
JEAN-PAUL SARTRE

In our country . . . one class of men makes war and leaves another to fight it out.
GENERAL WILLIAM SHERMAN

Boys are the cash of war. Whoever said we're not free-spenders doesn't know our likes.
JOHN CIARDI

War is the trade of kings.
JOHN DRYDEN

Sooner or later every war of trade becomes a war of blood.
EUGENE V. DEBS

Human beings do not fight for economic systems: who would be willing to die for capitalism? Certainly not the capitalists.
SIDNEY HOOK

When it comes time to hang the capitalists they will compete with each other to sell us the rope at a lower price.
VLADIMIR ILICH LENIN

War is capitalism with the gloves off.
TOM STOPPARD

The moral is obvious; it is that great armaments lead inevitably to war.
SIR EDWARD GREY

It is not armaments that cause war, but war that causes armaments.
SALVADOR DE MADARIAGA

A man may build himself a throne of bayonets, but he cannot sit on it.
WILLIAM RALPH INGE

We may find in the long run that tinned food is a deadlier weapon than the machine-gun.
GEORGE ORWELL

Beneath the rule of men entirely great
The pen is mightier than the sword.
EDWARD GEORGE BULWER-LYTTON

There are no manifestos like cannon and musketry.
ARTHUR WELLESLEY, DUKE OF
WELLINGTON

Guns before butter. Guns will make us powerful; butter will only make us fat.
> HERMANN GOERING

Fighting is essentially a masculine idea; a woman's weapon is her tongue.
> HERMIONE GINGOLD

Our swords shall play the orators for us.
> CHRISTOPHER MARLOWE

All they that take the sword shall perish with the sword.
> *Matthew 26:52*

(Armaments) constitute one of the most dangerous contributing causes of international suspicion and discord, and are calculated eventually to lead to war.
> CALVIN COOLIDGE

Well, boys, I reckon this is it: nuclear combat toe to toe, with the Rooskies!
> SLIM PICKENS, *as Major T.J. "King" Kong in* Dr. Strangelove or: How I Learned to Stop Worrying and Love the Bomb

The Russians can give you arms, but only the United States can give you a selection.
> ANWAR SADAT

There will one day spring from the brain of science a machine or force so fearful in its potentialities, so absolutely terrifying, that even man, the fighter, who will dare torture and death in order to inflict torture and death, will be appalled, and so abandon war forever. What man's mind can create, man's character can control.

THOMAS ALVA EDISON

Keep violence in the mind where it belongs.

BRIAN ALDISS

You can't say civilization don't advance . . . for every war they kill you a new way.

WILL ROGERS

It is a fundamental law of defense that you always have to use the most powerful weapon you can produce.

MAJOR GENERAL JAMES BURNS

We've got to take the atom bomb seriously. It's dynamite.

SAM GOLDWYN

The bomb that fell on Hiroshima fell on America, too.

HERMANN HAGEDORN

The atom bomb is a paper tiger with which the Americans try to frighten people.
MAO TSE-TUNG

The atom bomb will never go off, and I speak as an expert on explosives.
ADMIRAL WILLIAM LEAHY

These doomsday warriors look no more like soldiers than the soldiers of the Second World War looked like conquistadors. The more expert they become, the more they look like lab assistants at a small college.
ALISTAIR COOKE

The more horrible a depersonalized scientific mass war becomes, the more necessary it is to find universal ideal motives to justify it.
JOHN DEWEY

No country without an atom bomb could properly consider itself independent.
CHARLES DE GAULLE

The H-bomb rather favors small nations that don't as yet possess it; they feel slightly more free to jostle other nations, having discovered that a country can stick its tongue out quite far these days without provoking war, so horrible are war's consequences.
E.B. WHITE

No weapon has ever settled a moral problem. It can impose a solution but it cannot guarantee it to be a just one. You can wipe out your opponents. But if you do it unjustly you become eligible for being wiped out yourself.

ERNEST HEMINGWAY

Mark! where his carnage and his conquests cease!
He makes a solitude, and calls it—peace!

LORD BYRON

The Superpowers often behave like two heavily armed blind men feeling their way around a room, each believing himself in mortal peril from the other whom he assumes to have perfect vision.

HENRY KISSINGER

We may be likened to two scorpions in a battle, each capable of killing the other, but only at the risk of his own life.

J. ROBERT OPPENHEIMER

Nuclear superiority was very useful to us when we had it.

RICHARD M. NIXON

If we have an arms control agreement, the Russians will cheat. If we have an arms race, we will win.

GENERAL EARLE WHEELER

The way to win an atomic war is to make certain it never starts.

GENERAL OMAR BRADLEY

My God, we simply have to figure a way out of this situation. There's no point in talking about "winning" a nuclear war.

DWIGHT D. EISENHOWER

I do have to point out that everything that has been said and everything in their manuals indicates that, unlike us, the Soviet Union believes that a nuclear war is possible. And they believe it's winnable.

RONALD REAGAN

Nuclear disaster, spread by winds and waters and fear, could well engulf the great and the small, the rich and the poor, the committed and the uncommitted alike. Mankind must put an end to war or war will put an end to mankind.

JOHN F. KENNEDY

You can't have this kind of war. There just aren't enough bulldozers to scrape the bodies off the streets.

DWIGHT D. EISENHOWER

The terror of the atom age is not the violence of the new power but the speed of man's adjustment to it—the speed of his acceptance. Already bombproofing is on approximately the same level as mothproofing.

E.B. WHITE

The Atomic Age is here to stay—but are we?

BENNETT CERF

We are discussing the end of the world—or how to delay it.

STUART BLANCHE, *Archbishop of York*

The human race's prospects of survival were considerably better when we were defenseless against tigers than they are today when we have become defenseless against ourselves.

ARNOLD TOYNBEE

When nuclear dust has extinguished their betters,
Will the turtles surviving wear people-neck sweaters?

E.Y. "YIP" HARBURG

If World War III is fought with atom bombs the war after that will be fought with stones.

ALBERT EINSTEIN

The military mind always imagines that the next war will be on the same lines as the last. That never has been the case and never will be. One of the great factors in the next war will obviously be aircraft. The potentialities of aircraft attack on a large scale are almost incalculable.

MARSHAL FERDINAND FOCH

The power of an air force is terrific when there is nothing to oppose it.

WINSTON CHURCHILL

It is far more important to be able to hit the target than it is to haggle over who makes a weapon or who pulls a trigger.

DWIGHT D. EISENHOWER

The third peculiarity of aerial warfare was that it was at once enormously destructive and entirely indecisive.

H.G. WELLS

In the future, war will be waged essentially against the unarmed populations of the cities and great industrial centers.

GENERAL GIULIO DOUHET

I think it is well for the man in the street to realize that there is no power on earth that can protect him from being bombed. Whatever people may tell him, the bomber will always get through. . . . I just mention that . . . so that people may realize what is waiting for them when the next war comes.

STANLEY BALDWIN, *1932*

If civilians are going to be killed, I would rather have them be their civilians than our civilians.

STUART SYMINGTON

You've got to forget about this civilian. Whenever you drop bombs, you're going to hit civilians.
BARRY GOLDWATER

The civilian is a soldier on 11 months annual leave.
GENERAL YIGAEL YADIN

Well, we did not build those bombers to carry crushed rose petals.
GENERAL THOMAS S. POWER

The Bomb brought peace but man alone can keep that peace.
WINSTON CHURCHILL

The mere absence of war is not peace.
JOHN F. KENNEDY

Peace is a virtual, mute, sustained victory of potential powers against probable greeds.
PAUL VALÉRY

Generally speaking, it would be true to say that no one believes that war pays and nearly everyone believes that policies which lead inevitably to war do pay. Every nation sincerely desires peace; and all nations pursue courses which if persisted in, must make peace impossible.
SIR NORMAN ANGELL

I am not so senseless as to want war. We want peace and understanding, nothing else. We want to give our hand to our former enemies. . . . When has the German people ever broken its word?

ADOLF HITLER

It isn't enough to talk about peace. One must believe in it. And it isn't enough to believe in it. One must work at it.

ELEANOR ROOSEVELT

The world needs anger. The world often continues to allow evil because it isn't angry enough.

BEDE JARRETT

You can't separate peace from freedom because no one can be at peace unless he has his freedom.

MALCOLM X

Since wars begin in the minds of men, it is in the minds of men that the defenses of peace must be constructed.

Constitution of the United Nations Educational, Scientific and Cultural Organization

Peace cannot be kept by force. It can only be achieved by understanding.

ALBERT EINSTEIN

War is an invention of the human mind. The human mind can invent peace with justice.
NORMAN COUSINS

In the art of peace, Man is a bungler.
GEORGE BERNARD SHAW

The world will never have a lasting peace so long as men reserve for war the finest human qualities.
JOHN FOSTER DULLES

Long, continuous periods of peace and prosperity have always brought about the physical, mental and moral deterioration of the individual.
BRADLEY A. FISKE

God and the politicians willing, the United States can declare peace upon the world, and win it.
ELY CULBERTSON

The United States is not a country to which peace is necessary.
WILLIAM MCKINLEY

(Peace is) an idea which seems to have originated in Switzerland but has never caught hold in the United States. Supporters of this idea are frequently accused of being unpatriotic and trying to create civil disorder.
DICK GREGORY

Look at the Swiss! They have enjoyed peace for centuries and what have they produced? The cuckoo clock.

ORSON WELLES, *as Harry Lime in* The Third Man

The principle of neutrality . . . has increasingly become an obsolete conception, and, except under very special circumstances, it is an immoral and shortsighted conception.

JOHN FOSTER DULLES

There is such a thing as a man being too proud to fight.

WOODROW WILSON

He who walks in the middle of the road gets hit from both sides.

GEORGE P. SCHULTZ

I have given instructions that I be informed every time one of our soldiers is killed, even if it is in the middle of the night. When President Nasser leaves instructions that he is to be awakened in the middle of the night if an Egyptian soldier is killed, there will be peace.

GOLDA MEIR

I love peace, and I am anxious that we should give the world still another useful lesson, by showing to them other modes of punishing injuries than by war, which is as much a punishment to the punisher as to the sufferer.
THOMAS JEFFERSON

We will never be able to contribute to building a stable and creative world order until we first form some conception of it.
HENRY KISSINGER

There must be, not a balance of power, but a community of power; not organized rivalries but an organized common peace.
WOODROW WILSON

There is only one threat to world peace, the one that is presented by the internationalist communist conspiracy.
RICHARD M. NIXON

. . . they shall beat their swords into plowshares, and their spears into pruning-hooks: nation shall not lift up sword against nation, neither shall they learn war any more.
Isaiah 2:4

Let war yield to peace, laurels to paeans.
CICERO

We shall never stop war, whatever machinery we may devise, until we have learned to think always, with a sort of desperate urgency and an utter self-identification, of single human beings.
    VICTOR GOLLANCZ

Let us not deceive ourselves; we must elect world peace or world destruction.
    BERNARD M. BARUCH

I believe in compulsory cannibalism. If people were forced to eat what they killed there would be no more war.
    ABBIE HOFFMAN

There are many kinds of wars. One war has just ended but I do not know if peace has come.
INDIRA GANDHI

Don't tell me peace has broken out.
BERTOLT BRECHT

He that makes a good war makes a good peace.
GEORGE HERBERT

Those who win a war well can rarely make a good peace and those who could make a good peace would never have won the war.
WINSTON CHURCHILL

I say we are going to have peace even if we have to fight for it.
DWIGHT D. EISENHOWER

"Peace" is when nobody's shooting. A "just peace" is when our side gets what it wants.
BILL MAULDIN

It takes at least two to make peace but only one to make war.
NEVILLE CHAMBERLAIN

It takes two to make peace.
JOHN F. KENNEDY

When you're at war you think about a better life; when you're at peace you think about a more comfortable one.
THORNTON WILDER

Peace, n. In international affairs, a period of cheating between two periods of fighting.
AMBROSE BIERCE

Peace has her victories
No less renowned than war.
JOHN MILTON

If you want peace, understand war.
B.H. LIDDELL HART

Get in a tight spot in combat, and some guy will risk his ass to help you. Get in a tight spot in peacetime, and you go it all alone.
BRENDAN FRANCIS

Peace is not only better than war, but infinitely more arduous.
GEORGE BERNARD SHAW

Peace hath higher tests of manhood
Than battle ever knew.
JOHN GREENLEAF WHITTIER

The most disadvantageous peace is better than the most just war.
ERASMUS

Better to live in peace than to begin a war and lie dead.
CHIEF JOSEPH

There never was a good war, or a bad peace.
BENJAMIN FRANKLIN

A bad peace is better than a good war.
Russian proverb

A bad peace is even worse than war.
TACITUS

Everlasting peace will come to the world when the last man has slain the last but one.
ADOLF HITLER

Ours is a world of nuclear giants and ethical infants. We know more about war than we know about peace, more about killing than we know about living.
GENERAL OMAR BRADLEY

Our understanding of how to live—live with one another—is still far behind our knowledge of how to destroy one another.
LYNDON B. JOHNSON

Peace is always beautiful.
WALT WHITMAN

Either man is obsolete or war is.
BUCKMINSTER FULLER

We have met here to fight against war. The truth is that one may not and should not in any circumstances or under any pretext kill his fellow man.
LEO TOLSTOY

My pacifism is not based on any intellectual theory but on a deep antipathy to every form of cruelty and hatred.
ALBERT EINSTEIN

There are only two classes who, as categories, show courage in war—the front-line soldier and the conscientious objector.
B.H. LIDDELL HART

The pacifist is as surely a traitor to his country and to humanity as is the most brutal wrongdoer.
THEODORE ROOSEVELT

Pacifism is simply undisguised cowardice.
ADOLF HITLER

As a woman I can't go to war, and I refuse to send anyone else.
JEANNETTE RANKIN

An army of lovers shall not fail.
RITA MAE BROWN

Those who do not go to war roar like lions.
Kurdish proverb

Because I wore a peace symbol, I had to have an extra interview to determine my suitability as a member of the military.
SUSAN SHNALL

Sometime they'll give a war and nobody will come.
CARL SANDBURG

I discovered to my amazement that average men and women were delighted at the prospect of war. I had fondly imagined what most pacifists contended, that wars were forced upon a reluctant population by despotic and Machiavellian governments.
BERTRAND RUSSELL

A man who says that no patriot should attack the war until it is over is not worth answering intelligently; he is saying that no good son should warn his mother off a cliff until she has fallen over it.

G.K. CHESTERTON

What this country needs—what every country needs occasionally—is a good hard bloody war to revive the vice of patriotism on which its existence as a nation depends.

AMBROSE BIERCE

True patriotism hates injustice in its own land more than anywhere else.

CLARENCE DARROW

Each man must for himself alone decide what is right and what is wrong, which course is patriotic and which isn't. You cannot shirk this and be a man. To decide against your conviction is to be an unqualified and inexcusable traitor, both to yourself and to your country, let men label you as they may.

MARK TWAIN

You're not supposed to be so blind with patriotism that you can't face reality. Wrong is wrong no matter who does it or who says it.

MALCOLM X

What we need are critical lovers of America—patriots who express their faith in their country by working to improve it.

HUBERT HUMPHREY

Patriotism is your conviction that this country is superior to all other countries because you were born in it.

GEORGE BERNARD SHAW

The man who loves other countries as much as his own stands on a level with the man who loves other women as much as he loves his own wife.

THEODORE ROOSEVELT

Nationalism is an infantile disease. It is the measles of mankind.
ALBERT EINSTEIN

Patriotism is a kind of religion; it is the egg from which wars are hatched.
GUY DE MAUPASSANT

Patriotism is the willingness to kill and be killed for trivial reasons.
BERTRAND RUSSELL

Ubi libertas ibi patria,
Where liberty is, there is my country.
JAMES OTIS

Patriotism is the virtue of the vicious.
OSCAR WILDE

Patriotism is the last refuge of a scoundrel.
SAMUEL JOHNSON

Not for the flag
Of any land because myself was born there
Will I give up my life.
But I will love that land where man is free,
And that will I defend.
EDNA ST. VINCENT MILLAY

Ask not what your country can do for you—ask what you can do for your country.
JOHN F. KENNEDY

Liberty means responsibility. That is why most men dread it.
GEORGE BERNARD SHAW

The spirit of liberty is the spirit which is not too sure that it is right.
LEARNED HAND

One should never put on one's best trousers to go out to battle for freedom and truth.
HENRIK IBSEN

Man was born free and everywhere he is in shackles.
JEAN-JACQUES ROUSSEAU

The lamps are going out all over Europe; we shall not see them lit again in our lifetime.
SIR EDWARD GREY, *on the eve of World War I*

The tyrant dies and his rule is over; the martyr dies and his rule begins.
SØREN KIERKEGAARD

It is often pleasant to stone a martyr, no matter how much we admire him.
    JOHN BARTH

If the war didn't happen to kill you it was bound to start you thinking.
    GEORGE ORWELL

I only wish during the war they'd a took me in the army. I coulda been dead by now.
    ARTHUR MILLER

When it comes to dying for your country, it's better not to die at all.
    LEW AYRES, *as Paul Baumer in* All Quiet on the Western Front

Patriots always talk of dying for their country and never of killing for their country.
    BERTRAND RUSSELL

They died to save their country and they only saved the world.
    HILAIRE BELLOC

Whenever you hear a man speak of his love for his country it is a sign that he expects to be paid for it.
    H.L. MENCKEN

Crime isn't a mass instinct except in a time of war, and then it's merely an obscene sport.
S.S. VAN DINE

There's a consensus out there that it's OK to kill when your government decides who to kill. If you kill inside the country you get in trouble. If you kill outside the country, right time, right season, latest enemy, you get a medal.
JOAN BAEZ

It has been argued that, when killing is viewed as not only permissible but heroic behavior sanctioned by one's government or cause, the fine distinction between taking a human life and other forms of impermissible violence gets lost, and rape becomes an unfortunate but inevitable by-product of the necessary game called war.
SUSAN BROWNMILLER

Killing
Is the ultimate simplification of life.
HUGH MACDIARMID

Sometimes the worst thing we can know about a man is that he has survived. Those who say that life is worth living at any cost have already written for themselves an epitaph of infamy, for there is no cause and no person they will not betray to stay alive.
SIDNEY HOOK

. . . there are few that die well in a battle; for how can they charitably dispose of any thing when blood is their argument?
WILLIAM SHAKESPEARE, King Henry V

Dying is an art, like anything else.
SYLVIA PLATH

Death is an acquired trait.
WOODY ALLEN

The men that war does not kill it leaves completely transparent.
BRAZILIAN COLONEL CASTELO BRANCO

The question so often asked, "Would the survivors envy the dead?" may turn out to have a simple answer. No, they would be incapable of such feelings. They would not so much envy as, inwardly and outwardly, resemble the dead.

ROBERT JAY LIFTON

It takes twenty years or more of peace to make a man; it takes only twenty seconds of war to destroy him.

KING BAUDOUIN I, *of Belgium*

Wars, conflict, it's all business. One murder makes a villain. Millions a hero. Numbers sanctify.

CHARLIE CHAPLIN, *in the title role of* Monsieur Verdoux

A single death is a tragedy, a million deaths is a statistic.

JOSEPH STALIN

A dog run over by a car upsets our emotional balance. . . . Three million Jews killed in Poland causes us moderate uneasiness. Statistics don't bleed: it is the detail which counts.

ARTHUR KOESTLER

One murder makes a villain,
Millions, a hero.

BEILBY PORTEUS

What are the lives of a million men to me!
NAPOLEON BONAPARTE

The graveyards are full of indispensable men.
CHARLES DE GAULLE

Anybody who thinks that war is pleasant . . . you know, the old veteran stuff. You know, the "War is great stuff." Well, it's great for the survivors—not great for the people who are killed in it.
BERNARD B. FALL

I have already given two cousins to the war and I stand ready to sacrifice my wife's brother.
CHARLES FARRAR BROWNE (ARTEMUS WARD)

I was very careful to send Mr. Roosevelt every few days a statement of our casualties. I tried to keep before him all the time the casualty results because you get hardened to these things and you have to be very careful to keep them always in the forefront of your mind.
GENERAL GEORGE MARSHALL

The one thing I cannot forgive the Arabs for is that they forced our sons to kill their sons.
GOLDA MEIR

The time not to become a father is eighteen years before a war.
E.B. WHITE

Mother whose heart hung humble as a button
On the bright splendid shroud of your son,
Do not weep.
War is kind.
STEPHEN CRANE

If there must be trouble let it be in my day, that my
child may have peace.
THOMAS PAINE

In peace, sons bury their fathers; in war, fathers
bury their sons.
HERODOTUS

We fight for men and women whose poetry is not
yet written.
COLONEL ROBERT GOULD SHAW

The object of war is to survive it.
JOHN IRVING

I have always been against the pacifists during the
war, and against the jingoists at the end.
WINSTON CHURCHILL

Wars are different from baseball games where, at the
end of the game, the teams get dressed and leave
the park.
HARRY TRUMAN

Most people coming out of war feel lost and resentful. What had been minute-to-minute confrontation with yourself, your struggle with what courage you have against discomfort, at the least, and death at the other end, ties you to the people you have known in the war and makes for a time others seem alien and frivolous.

LILLIAN HELLMAN

A man who is good enough to shed his blood for his country is good enough to be given a square deal afterwards. More than that no man is entitled to, and less than that no man shall have.

THEODORE ROOSEVELT

The only war is the war you fought in. Every veteran knows that.

ALLAN KELLER

I was only good once—in a war. Some men should never come back from war.

JOHN DALL, *in* Another Part of the Forest

One of the main effects of war, after all, is that people are discouraged from being characters.

KURT VONNEGUT

The United States never lost a war or won a conference.

WILL ROGERS

They were never defeated, they were only killed.
Saying about the French Foreign Legion

Conquered people tend to be witty.
SAUL BELLOW

We have fought this fight as long, and as well as we know how. We have been defeated. For us, as a Christian people, there is now but one course to pursue. We must accept the situation.
ROBERT E. LEE

Too much success is not wholly desirable; an occasional beating is good for men—and nations.
ADMIRAL ALFRED THAYER MAHAN

There is nothing certain about war except that one side won't win.
SIR IAN HAMILTON

Victory at all costs, victory in spite of all terror, victory however long and hard the road may be; for without victory there is no survival.
WINSTON CHURCHILL

Victory has a thousand fathers, but defeat is an orphan.
JOHN F. KENNEDY

Who dares, wins.
Motto of the British Special Air Service regiment

There is no victory except through our imaginations.
DWIGHT D. EISENHOWER

Any coward can fight a battle when he's sure of winning.
GEORGE ELIOT

Morale is the greatest single factor in successful wars.
DWIGHT D. EISENHOWER

The will to conquer is the first condition of victory.
MARSHAL FERDINAND FOCH

A lost battle is a battle one thinks one has lost.
MARSHAL FERDINAND FOCH

There is only one decisive victory: the last.
KARL VON CLAUSEWITZ

In starting and waging a war, it is not fight that matters but victory.
ADOLF HITLER

In war there is no second prize for the runner-up.
GENERAL OMAR BRADLEY

In war there is no substitute for victory.
DWIGHT D. EISENHOWER

When you are winning a war almost everything that happens can be claimed to be right and wise.
WINSTON CHURCHILL

The victor will not be asked afterwards whether he told the truth or not.
ADOLF HITLER

Only the winners decide what were war crimes.
GARY WILLS

People never lie so much as after a hunt, during a war or before an election.
OTTO VON BISMARCK

If you live long enough, you'll see that every victory turns into a defeat.
SIMONE DE BEAUVOIR

One more such victory and we are undone.
PYRRHUS, *of Epirus*

In war, whichever side may call itself the victor, there are no winners but all are losers.
NEVILLE CHAMBERLAIN

Nothing except a battle lost can be half so melancholy as a battle won.
ARTHUR WELLESLEY, DUKE OF
WELLINGTON

The quickest way of ending a war is to lose it.
GEORGE ORWELL

You can no more win a war than you can win an earthquake.
JEANNETTE RANKIN

In war, Resolution; in defeat, Defiance; in victory, Magnanimity; in peace, Goodwill.
WINSTON CHURCHILL

INDEX

INDEX

INDEX

INDEX